GROWING THE AUTOMOTIVE SUPPLY CHAIN

Dr Matthias Holweg
Judge Business School
University of Cambridge

with
Yung Tran
Society of Motor
Manufacturers and Traders

Philip Davies
Department for
Business, Innovation and Skills

Stephan Schramm
Judge Business School
University of Cambridge

Foreword by:
Vince Cable
Secretary of State,
Department for Business, Innovation & Skills

Published by:
PICSIE Books
Box 622
Buckingham, MK18 7YE
United Kingdom

How to order:
PICSIE Books
Telephone & Fax: +44 (0) 1280 815 023
Web site: www.picsie.co.uk
E-mail: picsiebook@btinternet.com

Copyright © PICSIE Books, 2011
ISBN 10 0 9541244-9-9
ISBN 13 978-0-9541244-9-6
EAN 978095954124496
British Library Cataloguing-in-Publication Data
A catalogue record for this book is available from the British Library

Cover photograph courtesy of Aston Martin.

FOREWORD

Companies in the UK are already winning more than £7 billion of work annually from our vehicle makers. But it is clear there are opportunities to grow this business, taking advantage of UK innovation, engineering skills and the logistical benefits of 'local' supply.

On the back of the work being done by the Automotive Council for this report, more than £100 million of manufacturing work has now come back to suppliers in the UK, safeguarding and creating vital jobs.

The automotive sector is hugely important to the UK and is our number one manufactured export, with over 300,000 manufacturing jobs accounting for 12 per cent of the UK's total manufacturing employment and a further 480,000 in the motor retail sector.

The Department for Business, Innovation & Skills is leading the Coalition Government's Growth Review. Our goal over the next 10 years is to grow manufacturing in the UK. Also, to make the UK Europe's largest exporter of high-value goods and related services. And, to increase the number of people taking up manufacturing as a career.

These are challenging ambitions, but I know the Automotive Council is working across this agenda to enhance the attractiveness of the UK as a location for global automotive investment, to promote UK-based manufacturers and technologies, to strengthen the supply chain, and position the UK as a leading global player in developing low and ultra-low carbon vehicles and technologies.

This report typifies what can be achieved when government and industry work in partnership and is an important step in the Government's growth agenda. I am most grateful to all the members for their contributions and hard work.

Vince Cable

Secretary of State,
Department for Business, Innovation & Skills
Co-Chair, Automotive Council

EXECUTIVE SUMMARY

Objective

Growing the UK auto supply chain is seen as an issue of the highest priority by the Automotive Council. This 'sourcing roadmap' provides an overview of current and prospective sourcing patterns in the UK automotive industry. It serves as the empirical grounding for determining and prioritizing activities by the Automotive Council to retain and build supply chain capabilities in the UK automotive industry.

Key Findings

The UK automotive supply chain largely supports the vehicle programmes assembled in the UK; at present, about 80% of all component types required for vehicle assembly operations can be produced by UK suppliers.

The combined UK purchasing spend of the UK-based automotive, commercial vehicle and yellow goods manufacturers is £7.4bn per annum. The amount purchased in the UK equates to c.36% of their global purchasing spend.

In terms of value-added at supplier level, on average 74% of UK-based suppliers manufacture in the UK (as opposed to those who assemble, late configure, or distribute components); this ratio is lower at first tier suppliers where 65% manufacture, while virtually all second-tier suppliers operate manufacturing facilities in the UK. The average supplier serves six customers (median), with a strong bias towards those OEMs that operate vehicle and engine assembly plants in the UK.

'Proximity' was identified as the key competitive advantage of UK suppliers: in operational terms, proximity allows for (1) lower logistics cost, a better support for UK-built vehicles, (2) the responsive configuration of parts, as well as (3) for more flexibility to adjust to volume and product mix fluctuations. In strategic terms proximity also acts as (4), a general proxy for risk reduction the supply chain, as well as (5), a hedge against currency fluctuations.

The key reason why UK suppliers have lost business is that their unit cost was not competitive. The OEMs further state that secondary reasons for not sourcing from UK suppliers were, in order of prevalence, (1) the general lack of accredited suppliers, (2) required processing capabilities that were not available, (3) quality and (4) logistics that were not competitive. While it is not possible to identify any specific patterns here, the main reasons however can be summarised as (1) operational execution (QCD), as well as (2), financial aspects (availability of finance, concerns over supplier size and stability).

When UK supply business is lost, about one third stays within the UK and Western Europe, one third goes to low-cost countries, while the final third show no clear pattern. The risk of losing business to low-cost regions rises however considerably for second-tier suppliers. Nonetheless, there exists a sizeable opportunity to win significant fractions of this business back into the UK.

Opportunities

Several key opportunities for retaining and building supply chain capabilities in the UK automotive industry could be identified from the surveys and analyses. These have been assembled into a 'UK sourcing roadmap' (also see p/40 for a large-scale version), which visualises the key short-term and medium-term opportunities, as well as the areas where critical support is required to support the UK automotive supply chain.

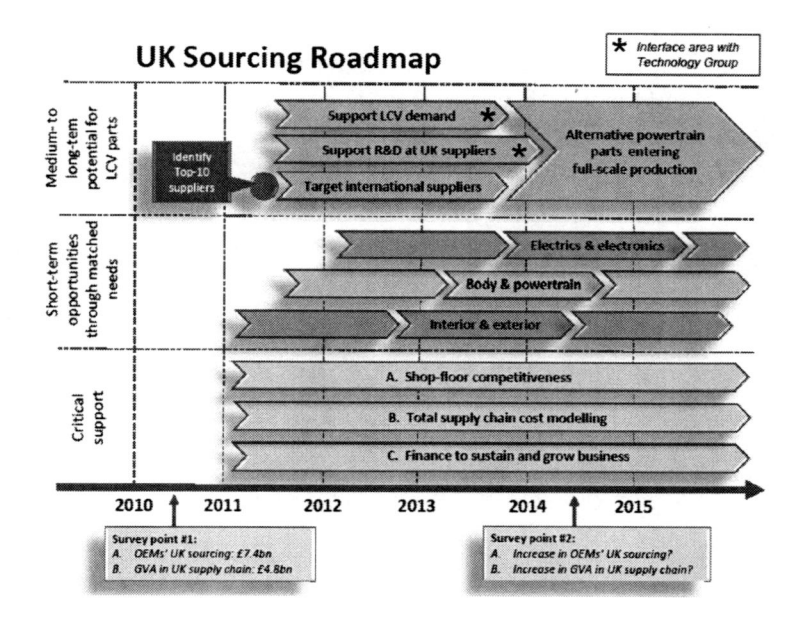

UK Sourcing Roadmap

★ Interface area with Technology Group

Medium- to long-term potential for LCV parts:
- Identify Top-10 suppliers
- Support LCV demand ★
- Support R&D at UK suppliers ★
- Target international suppliers
- Alternative powertrain parts entering full-scale production

Short-term opportunities through matched needs:
- Electrics & electronics
- Body & powertrain
- Interior & exterior

Critical support:
- A. Shop-floor competitiveness
- B. Total supply chain cost modelling
- C. Finance to sustain and grow business

Timeline: 2010 — 2011 — 2012 — 2013 — 2014 — 2015

Survey point #1:
A. OEMs' UK sourcing: £7.4bn
B. GVA in UK supply chain: £4.8bn

Survey point #2:
A. Increase in OEMs' UK sourcing?
B. Increase in GVA in UK supply chain?

Short term opportunities largely arise where OEM sourcing needs match with strategic growth areas of UK suppliers, in other words, where OEMs have a current need that could potentially be met from a UK-based supplier. Here, considerable potential was identified for the 'classic' components sourced:

1. 'Powertrain & body' components, where virtually all needs can theoretically be met, with the exception of 'heavy metals' processing capabilities (casting, forging, etc) that were identified as supply chain constraint by both OEM and suppliers.
2. 'Interior and exterior' components, where virtually all OEM needs can be matched with supplier growth intentions.
3. 'Electrics & electronics' components, where some needs can be matched. Critical parts missing are batteries, and electronics in general.

A comprehensive list of commodities that OEMs wish to procure was developed as a guide to identify and foster specific supply opportunities.

Medium- to long term opportunities for building the UK supply chain arise from the gradual shift towards a portfolio of powertrain architectures, and the novel components these will require. At present, UK suppliers do not feel they are at the forefront of supporting this shift. Here, the 'Top-10' most desirable low carbon powertrain suppliers will be identified, in order to target efforts to entice these to commence operation in the UK.

Finally, in terms of further critical support areas, the surveys show that UK suppliers are losing out on a unit cost basis, while – paradoxically – proximity was rated as the most important competitive factor. Hence a key opportunity arises to help suppliers conceptually to make their business case on a 'total supply chain cost' basis, rather than on a unit cost basis alone. A total cost model, aimed at helping UK suppliers to make a more comprehensive business case, is developed and presented in the appendix of this report.

0. ABOUT THIS REPORT

Based on the findings of the New Automotive Innovation and Growth Team (NAIGT), the Automotive Council has identified strengthening the UK auto supply chain as an issue of the highest priority. A key first step towards this objective will be the production of a 'sourcing road map', which will provide an overview of current and prospective sourcing patterns in the UK automotive industry.

The main objective is to generate actionable insights to support industrial policy by:

1. assessing the scope of decision making ('Who decides on sourcing?')

2. understanding the UK automotive supply chain's relative competitiveness ('Why is UK supply business lost, and where is it lost to?')

3. identifying potential for retaining /building supply chain capabilities ('What is on the manufacturers' sourcing wish list, where do suppliers envisage growth?')

The main thrust of this research is to identify areas of concerns, as well as clusters of opportunity to sustain and grow automotive sourcing from UK suppliers. The report provides an impartial, independent analysis from both a manufacturer and a supplier point of view. To this effect we have conducted surveys of the purchasing directors of all UK vehicle manufacturers (OEMs), as well as conducted a survey of first and second tier automotive suppliers in the UK. This data was complemented with secondary data, where appropriate.

The report is structured as follows: Section I will provide a brief economic overview of the UK automotive supply chain sector; Section II will outline the methodology used to develop the 'sourcing roadmap'. Sections III and IV will present the key findings from the OEM and supplier surveys, respectively. Section V summarises the key findings and presents the 'UK Sourcing Roadmap'.

Appendix A provides a list of components, by category, where OEM requirements, could be matched with supplier growth areas.

Appendix B provides an overview of current OEM sourcing patterns, by component category.

Appendix C introduces a *total supply chain cost* model that was developed in order to support the business case when UK suppliers bid for new business.

I. THE UK AUTO SUPPLY CHAIN

Supply Chain Overview

The supply chain is crucial to the automotive industry, representing about 40% of the retail price of a passenger car. In other words, vehicle manufacturers buy in about 60-75% of the value from the component supply chain[i]. The cost of materials and parts is around six times the cost of final vehicle assembly. It is estimated that every job in vehicle assembly supports 7.5 elsewhere in the economy[ii], with the process of making the parts somewhat more labour-intensive than final assembly, much of which is heavily automated.

Around 2,350 UK companies regard themselves as 'automotive' suppliers[iii], with latest data showing that they employed around 82,000 people at the end of 2009[iv]. This number however should be treated with some caution, as component suppliers could classify themselves either via the industry they predominantly serve (e.g. "automotive"), via their production process (e.g. "pressing"), or via their main products (e.g. "seats"). Only in the first case would the official statistics classify the suppliers as "automotive", while in the latter two cases the likely classification would under "manufacturing" in general. In that sense the official classifications can potentially be misleading. Unfortunately there are no additional data sources at hand that could be used to triangulate this data.

The table and chart below shows the evolution of the number of automotive supply chain businesses in recent years, and how this breaks down into the four SIC codes[v] that are specific to the automotive sector (Source of data: ONS ABI 2009, the latest available).

A period of modest growth peaked in 2005, with the number of supply chain businesses falling away slightly through 2007/8, before a sharp fall through the recession where nearly 10% of UK supply chain businesses closed, as sales fell 25% and value added was one third down. This indicates that the sector has lost considerable capacity through the recession that cannot easily be replaced as business turns up.

	SIC code	1997	1998	1999	2000	2001	2002	2003	2004	2005	2006	2007	2008	2009
Bodies	34.2 / 29.2	767	764	777	779	803	800	831	876	911	892	902	895	855
Parts	34.3 / 29.32	1,359	1,392	1,449	1,437	1,467	1,487	1,470	1,474	1,486	1,421	1,424	1,472	1,264
Tyres	25.11 / 22.11	124	145	131	148	167	166	161	147	130	126	116	98	78
Electrical	31.61 / 29.31	192	209	215	226	227	229	233	233	248	250	251	201	161
Total		2,442	2,510	2,572	2,590	2,664	2,682	2,695	2,730	2,775	2,689	2,693	2,666	2,358

The development of the total number of companies in the UK automotive supply chain (all SIC codes) can also be shown graphically:

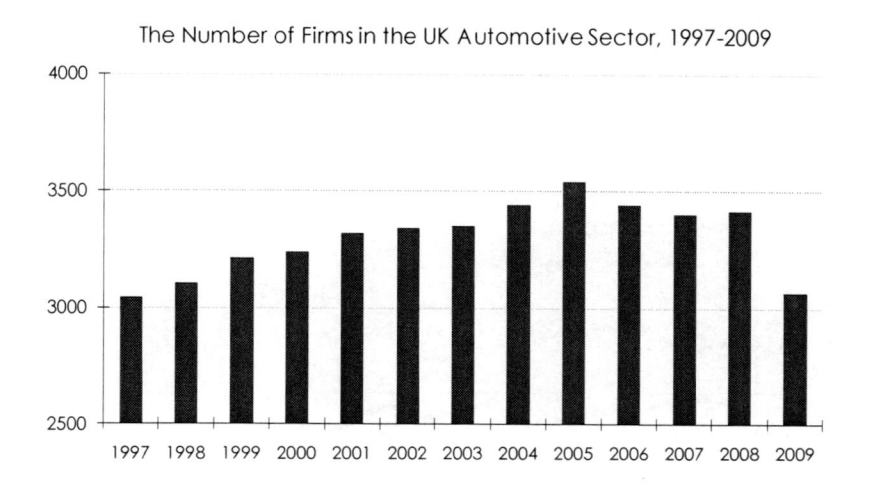

The Number of Firms in the UK Automotive Sector, 1997-2009

Through the trough of the recession in 2009, the UK automotive supply chain generated around £3.1 billion of added value on sales of £12 billion[vi]. This compares with a 10-year period when value added was in the order of £4.5 to £5bn annually.

In terms of Gross Value-Added[vii], the automotive sector contributes an average £9.3bn (1995-2009 average) to the UK economy. This equates to between 5-7% of the overall contribution of the total manufacturing sector.

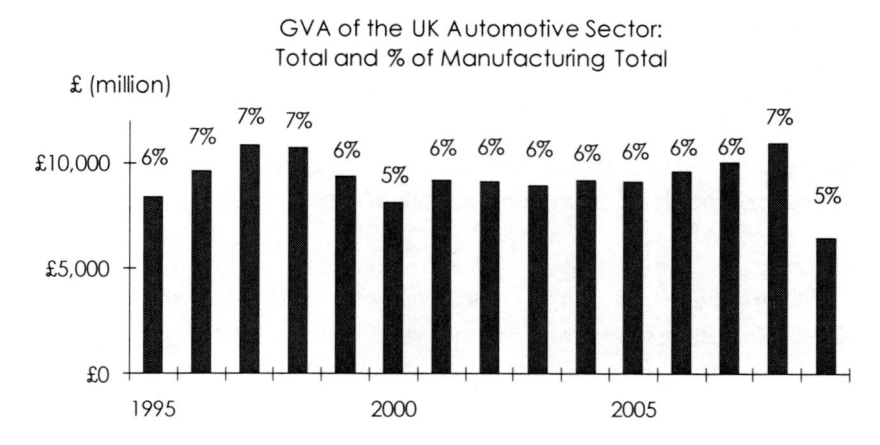

GVA of the UK Automotive Sector: Total and % of Manufacturing Total

Parts sector exports have been fairly flat at a little over £6 billion-worth of goods annually from the mid-1990's, though imports have grown from a similar level, to nearly £15 billion during the peak just before the recession, yielding a deficit of over £8bn at the peak. As markets have recovered, parts imports are again rising more rapidly than exports, possibly in part due to the loss of capacity noted above, and by mid-2010 the parts trade deficit was approaching £7bn on an annualised basis[viii].

UK Automotive Parts Trade

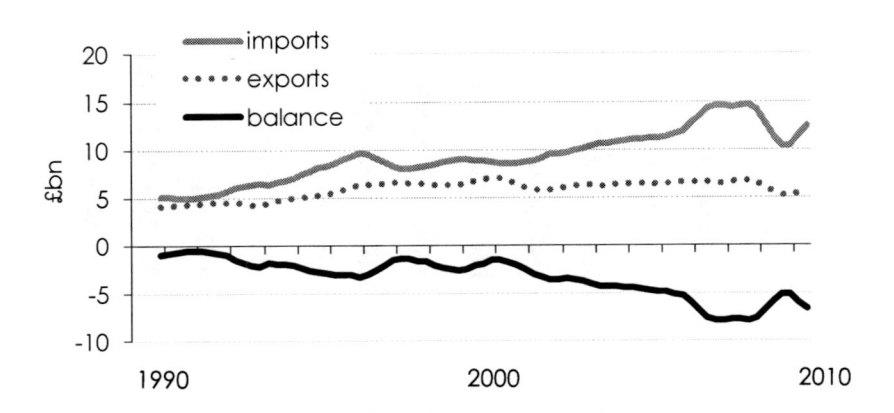

The map below shows the regional disposition of the automotive industry. It shows the location of the main vehicle manufacturers (including heavy goods vehicles, buses and construction equipment), as well as the location of the automotive supply base in the UK. This does not include companies in the wider manufacturing sector, or service suppliers, though the geographical disposition is likely to be similar.

This is only a partial view of the supply chain, as around two thirds of vehicle makers' inputs are from elsewhere in the economy, so the true size of the UK automotive supply chain is probably nearer 250,000 people. The wider supply chain includes raw materials, metal pressings, forgings and castings, glass, textiles, plastics, electronics and a wide variety of other manufactured products. It also includes services such as catering and security, all the agency workers, and also companies contracted to work in other areas such as materials handling and transport.

The majority of Companies operating in the automotive sector are small and medium sized enterprises. Of 2,900 businesses in the automotive sector for which data is available, just 80 have more than 250 employees, whilst nearly 2,000 businesses have less than 10 employees[ix].

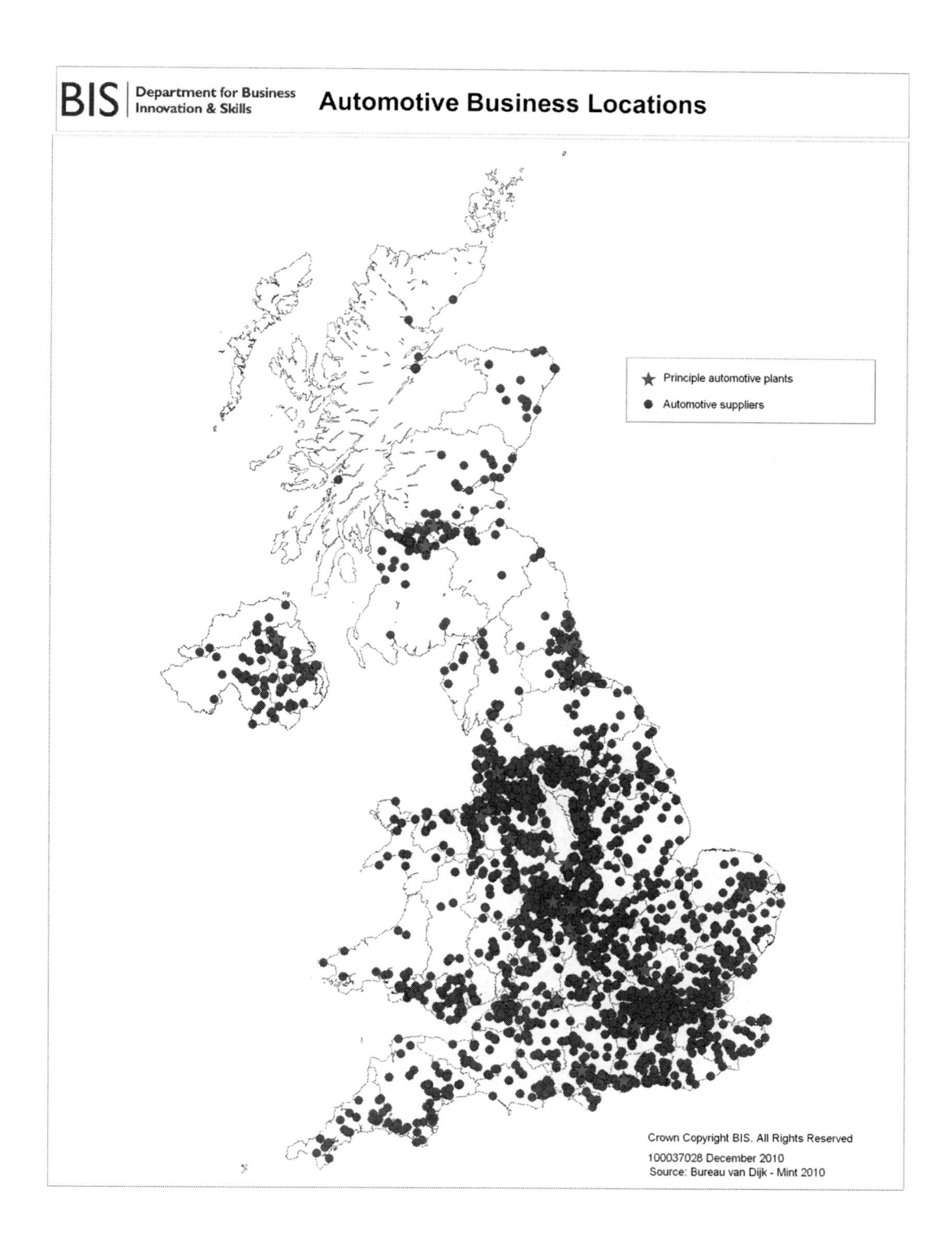

Automotive Business Locations

Principle automotive plants
Automotive suppliers

Crown Copyright BIS. All Rights Reserved
100037028 December 2010
Source: Bureau van Dijk - Mint 2010

II. A SOURCING 'ROADMAP'

A 'technology roadmap' depicts shift over time, generally from one state to another. Most common applications include technology shifts (e.g. in ICT, or automotive powertrains). Applied to sourcing, this raises conceptual problems, as it is dynamic & ongoing. One could argue that alternative low carbon powertrain parts will gradually phase-in, however, these will be in addition to traditional parts, so the shift is only partial.

We thus use the term 'road map' here in a loose sense to provide an overview of:

1. those components that currently can be sourced in the UK,
2. those components where there is a currently untapped potential for UK suppliers to provide such components,
3. components where there is a near miss in cost, quality, or other, where over the medium term the UK can develop solutions to address this gap,
4. those where there is no UK capability at present, yet the UK could consider actions to attract suppliers to fill the capability gap, and
5. new powertrain technologies and related components where collaboration and investment is required to ensure the UK can exploit such future opportunities.

Steps in producing the UK sourcing road map

In a first step, the purchasing directors of the major UK-based vehicle manufacturers (passenger cars and commercial vehicles) have been surveyed in order to identify which components are currently sourced in the UK, which ones are not, and why.

Based on these findings, a supplier questionnaire was developed to cross-validate the findings from Step I, by assessing the reciprocal views of the UK-based automotive suppliers. Combined, the results of I. and II. allow for the identification which component groups provide the best match – current and prospective - between OEM requirements and UK supply capabilities.

An event will be used as the main forum to present the results to industry, and encourage dialogue between suppliers and customers. This should be free to all participants, with high-level representation of OEM purchasing directors being key to the success of this event.

Further research may include an R&D capability review in conjunction with the Technology subgroup of the Automotive Council.

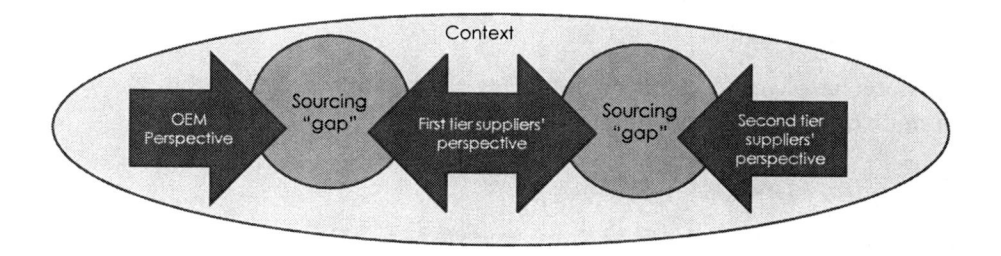

Confidentiality

The OEM and supplier surveys were conducted by the University of Cambridge, on behalf of the Automotive Council and its Supply Chain Group. As the data in question is of a commercially sensitive nature, only fully anonymised results will be presented.

For more detail on the methodology and survey design used please feel free to contact Dr Matthias Holweg at the Judge Business School, University of Cambridge, by email: m.holweg@jbs.cam.ac.uk.

III. A SURVEY OF UK PURCHASING DIRECTORS

In this section we will report on the findings of the survey of purchasing directors at the UK vehicle manufacturers, including the main commercial vehicle and yellow goods producers.

Overall we surveyed 11 purchasing directors, a sample which includes all seven volume car manufacturers that are producing in the UK, as well as the key truck and construction equipment producers. We thus have a full representation of the population of the UK car manufacturers, and hence, by definition also have a representative sample.

In order to convert a vehicle into feasible groupings, we split the 2,000 - 4,000 components in a car into 7 main categories:

1. Powertrain
2. Chassis
3. Interior
4. Exterior
5. Electrics
6. Electronics
7. Consumables

We will report our findings along the lines of these categories.

CURRENT UK SOURCING

The key question here was to identify what OEMs are currently sourcing from the UK, and who made the decision to source from here.

Purchasing Spend in the UK

The total purchasing spend of the 11 firms surveyed is £7,416 million, which accounts for a mean of 36% of their global purchasing spend. In other words, about one third of the value of components needed to support UK-based vehicle production is currently purchased in the UK, while two thirds are imported (please also refer to the parts trade import-output balance shown in Section for more detail).

The fraction of the amount purchased in the UK, in relation to their worldwide purchasing spend, ranges from 11% to 83% for the individual companies.

In terms of sourcing, Components account for an average of 65%, raw materials for 6%, while manufacturing-related services account for 10%. This spread is to be expected, considering that between 40-50% of all value-added is created in the component supply chain.

Answer	Average Value	Standard Deviation
Components	65.3%	23.2
Raw materials	6.4%	5.5
Manufacturing-related services	10.0%	10.8
Indirect suppliers	16.3%	20.3
Other	2.0%	4.5
Total	100.0%	

Scope of decision making

In order to assess the scope for retaining UK automotive supply business, it is important to understand where the OEM sourcing decisions are made, in order to target efforts appropriately.

Question	Strongly Disagree	Somewhat Disagree	Ambivalent	Somewhat Agree	Strongly Agree	N	Mean	CI[1] (±)
The sourcing decisions are unique to each assembly plant	4	3	1	2	1	11	-0.64	0.85
The decision where to source from rests within our UK operation	5	1	0	2	3	11	-0.27	1.09
Our sourcing strategy is determined for the entire group, not just for a single country or region	0	1	0	5	5	11	1.27	0.53
We have to source from preferred suppliers prescribed by our headquarters	3	4	0	4	0	11	-0.55	0.76
Our sourcing process is open to new suppliers	0	1	0	5	5	11	1.27	0.53
By default, we tend to look for suppliers from within our pool of established suppliers	0	0	1	10	0	11	0.91	0.18
Currency exchange rates have a strong influence on our sourcing strategy, even in the short-term	3	1	1	5	1	11	0.00	0.87
Unit cost is the key deciding factor in whether or not a supplier contract is awarded	1	3	0	6	1	11	0.27	0.75

[1] – 95% confidence interval

As the table above shows, all UK operations surveys have a certain scope of influence in terms of decision-making, however, it is also obvious that sourcing decisions are made considering group level implications. Overall we observe a bi-polar response on the 'centre of gravity' where final purchasing decision is made: for one part this is on the UK, for other OEMs this is outside the UK. Nonetheless all responses indicate that the UK operations have the ability to influence this decision.

In terms of new suppliers, we found a surprisingly small window of opportunity for new suppliers. On average, UK-based OEMs will start working with 4% new suppliers each year, which reflects a range of 0.5% to 10%. There is also a strong bias towards existing, large suppliers with global footprint. (Note: the ratio of new suppliers refers to both UK-based suppliers, as well as overseas firms).

One reason - frequently mentioned - is that there are not sufficient numbers of 'accredited suppliers'. The accreditation levels here include OEM-specific programmes, as well as generic ones such as ISO/TS16949:2002, ISO9000 or ISO14001.

When the business case is made for a new sourcing arrangement, the following variables are considered in addition to overall unit cost.

Answer	Response	%
Logistics cost to assembly plant	11	100%
Labour cost	8	73%
Taxes and tariffs	8	73%
Cost related to quality control	7	64%
Currency risk	6	55%
Other (see below)	4	36%
Labour cost inflation	4	36%
Transportation cost inflation/volatility	4	36%

The information consulted when looking for a new supplier includes primarily internal and external directories (67% each), while regional development agencies and supplier fairs play minor roles only. Buyer experience was mentioned as a further common point, which indicates a strong tendency of path dependency (i.e. sticking with established existing suppliers, rather than seeking new ones).

Overview of current sourcing from the UK

The overview of UK component sourcing (ranked by share of UK sourced) is shown below. This table is an important starting point for defining the "Sourcing Roadmap", as it shows the degree to which firms currently source component categories in the UK.

Components	UK sourced? Index 0-1	Single source? Index 0-1	In-house supplier? Index 0-1	Approx. annual volumes [units]	Expected volume change? Index -2 to +2
	Mean	Mean	Mean	Mean	Mean
Components currently sourced in the UK:					
Fuel tanks	0.91	0.45	0.09	248,667	+ 0.45
Engine components	0.80	0.30	0.00	284,500	+ 0.55
Headliners	0.73	0.36	0.00	248,500	+ 0.45
Adhesives and sealers	0.73	0.30	0.00	378,667	+ 0.45
Small plastics parts/fasteners	0.73	0.27	0.00	10,252,200	+ 0.45
Interior trim	0.73	0.09	0.00	1,718,833	+ 0.45
Spoilers and body cladding	0.70	0.30	0.00	117,000	+ 0.50
Carpets	0.64	0.45	0.00	3,381,833	+ 0.45
Seats	0.64	0.36	0.13	918,833	+ 0.45
Bumpers	0.60	0.56	0.30	4,046,167	+ 0.56
Panels	0.60	0.22	0.29	898,714	+ 0.56
Castings	0.56	0.22	0.11	1,504,600	+ 0.50
Instrument panels	0.55	0.50	0.22	248,500	+ 0.50
Glass	0.50	0.20	0.00	1,227,833	+ 0.40
Nuts, bolts, screws	0.50	0.20	0.00	20,252,200	+ 0.50
HVAC units	0.45	0.55	0.09	248,000	+ 0.45
Suspensions (struts)	0.27	0.36	0.10	881,167	+ 0.45
Forgings	0.22	0.33	0.00	1,104,600	+ 0.60
Engine control unit (ECU)	0.20	0.50	0.11	248,500	+ 0.50
Transmission components	0.20	0.30	0.13	294,500	+ 0.45
Power steering	0.18	0.27	0.10	244,500	+ 0.36
Tyres	0.18	0.00	0.00	895,500	+ 0.45
Alternators	0.09	0.45	0.00	248,500	+ 0.45
Components currently not sourced in the UK:					
Anti-lock brakes (ABS)	0.00	0.40	0.00	877,833	+ 0.40
Wheels	0.00	0.36	0.00	895,500	+ 0.45
Harnesses	0.00	0.36	0.00	1,734,833	+ 0.45
Entertainment / radio systems	0.00	0.33	0.00	247,167	+ 0.50
Brakes (discs, drums)	0.00	0.27	0.00	886,833	+ 0.45
Batteries	0.00	0.27	0.00	249,500	+ 0.45

The above data shows the level of 'UK sourced' as a 0-1 binary variable, which 1=UK sourced (at least in parts), and 0=not sourced from the UK. The score shown here is the average across all firms surveyed. The score for the other variables are calculated in the same way. (For full details for each of the seven categories see Appendix B).

This table provides the initial starting point towards defining the main opportunities for the UK economy as a whole: here, one could argue that the component categories currently not sourced in the UK provide the greatest opportunity for growth. This is not true, for the simple reason that in these areas no supplier capability is present. In that sense, offering these components out of the UK would require attracting new supplier investment to the UK, rather than expanding on an existing capability. For reasons further discussed below, attracting new suppliers to the UK is bound to be considerably more challenging, than working with existing UK suppliers and vehicle manufacturers on increasing their local sourcing.

In that sense the greatest opportunities, in our view, present themselves around those components that currently are (partly) sourced from the UK, where the vehicle manufacturers envisage growth, and which shows significant annual volumes overall.

Thus, the main categories that present opportunities are:

1. 'Interior and exterior' parts, which include interior parts such headliners, carpets, seats, Instrument panels and HVAC units, glass, as well exterior parts: bumpers, and large plastic mouldings,

2. 'Powertrain & body' parts, including suspensions, castings, forgings, and transmission components,

3. 'Electrics and electronic' parts, which includes power steering, engine control units, and alternators.

Reasons for sourcing from the UK

The following two questions assess the competitiveness of the UK supply chain. A key question is why OEMs source from the UK, i.e. what specific factors they see as providing the UK with its competitive advantage. Here, the strongest features that could be identified are related to proximity between the OEM assembly plant, and the supplier through the lowest logistics cost, or the vehicle only being made in the UK. We see some variation by component group, but these differences are not significant statistically.

The question underlying the table below is: 'For those component categories that are currently sourced from the UK, why was this source chosen?'

Components	Lowest logistics cost	Need for configuration close to the plant	Lowest unit cost	Unique supplier capabilities	Highest quality	The vehicle is only made in the UK	It is company policy to source in the UK	Other
Powertrain	7	3	6	3	4	3	1	
Exterior	8	8	4	3	4	3	1	
Interior	8	9	5	2	3	5	2	
Chassis	5	2	5	0	2	2	1	in-house
Electronics	0	0	2	1	1	1	1	
Electrical	0	1	1	1	0	0	1	
Consumables	4	4	4	0	1	1	0	
Total	32	27	27	10	15	15	7	1

Reasons for *not* sourcing from the UK

In many ways the more interesting question is why certain component categories are not sourced from the UK. Here, we asked the OEM Purchasing Directors to outline the key reasons, by component group:

Components	Unit cost was not competitive	Quality was not competitive	Delivery/ logistics was not competitive	Volumes/ supplier capacity was insufficient	Lack of accredited suppliers in the UK	The sourcing decision was taken abroad	Required raw materials not available	Required materials processing capabilities not available
Powertrain	4	1	2	1	5	0	1	3
Exterior	5	2	3	0	3	0	0	1
Interior	3	2	2	0	2	0	0	1
Chassis	5	2	2	1	5	1	2	4
Electronics	7	3	2	3	8	1	1	3
Electrical	7	4	1	3	8	1	1	3
Consumables	3	0	1	0	1	1	1	2
Total	34	14	13	8	32	4	6	17

As can be seen, uncompetitive unit cost is the main issue, as is the lack of accredited suppliers. This finding is in many ways not very helpful, as it provides little insight into the root cause behind this failure. This issue was picked up again in the component supplier survey, to match the findings from both views in the supply chain.

Outlook of purchasing spend

In terms of outlook, there is an overall predicted slight increase of UK sourcing, however, this finding is weak overall and not unanimous across firms. Furthermore, it is interesting to note that current UK suppliers do not seem to submit competitive bids, which confirms the earlier findings.

Question	Strongly Disagree	Slightly Disagree	No opinion either way	Somewhat Agree	Strongly Agree	N	Mean
Our UK-sourced purchasing volume is likely to increase in the next five years	0	3	1	5	2	11	0.55
It is our strategic intent to increase sourcing from within the UK	0	2	1	6	2	11	0.73
We have little influence over supplier selection as this is done outside of the UK	5	4	0	1	1	11	-1.00
Our UK suppliers are largely subsidiaries of global supplier groups	0	1	2	3	5	11	1.09
When put out a call for a new component, we receive competitive bids from UK suppliers	0	5	4	2	0	11	-0.27
We largely source from suppliers with a global footprint that can serve all our manufacturing facilities, not just the UK	1	2	1	5	2	11	0.45

POTENTIAL UK SOURCING

In this part of the survey we assess the potential for increasing the sourcing from UK component suppliers. Specifically, we asked about the parts OEMs would like to source in the UK, but that currently are not available. This 'wish-list' is a key component of the UK Sourcing Roadmap. These detailed requirements are listed in Section V 'The Sourcing Roadmap', and will not be repeated here.

Pooled Purchasing Volumes

Another issue sometimes mentioned is whether UK purchasing volumes are too low to attract investment. Clearly this is not necessarily always the case as the UK continues to win new investments, including in the supply chain. However, if one considers *total* (as opposed to individual OEM) purchasing requirements in the UK for certain products, the scale of this combined or pooled requirement could help attract suppliers into the UK.

The question we put to the OEMs was "Within the realms of the competition regulation, which parts would you – in principle - consider suitable for a pooled sourcing arrangement, to be shared with other OEMs?", the answer to which is shown below.

	'Never'	'Unlikely'	'Maybe'	'Likely'	'Definitely'	N	Mean	CI[1] (±)
Powertrain	1	2	3	4	1	11	0.18	0.69
Exterior	2	2	3	3	1	11	-0.09	0.77
Interior	2	3	1	5	0	11	-0.18	0.74
Chassis	1	1	5	2	2	11	0.27	0.70
Electronics	1	5	1	3	1	11	-0.18	0.74
Electrical	1	4	2	3	1	11	-0.09	0.72
Consumables	0	0	5	2	4	11	0.91	0.56
Raw Materials	1	1	2	3	4	11	0.73	0.80
Energy	1	1	3	4	2	11	0.45	0.72

[1] – 95% confidence interval

As the findings above show, overall there seems considerable scope to make the case that total UK purchasing volumes and requirements for certain products are much larger than suppliers might imagine, and that bringing this evidence to their attention when promoting the case for new UK investment would be worthwhile. However, as the data above reflects, there are many hurdles to such 'pooling'. Most prominent concerns include the willingness of OEMs to share purchasing data (for both competitive advantage and legal (competition) reasons). Overall this issue is likely to remain a difficult one.

SECONDARY SOURCING

This part of the survey goes beyond the traditional 'component' sourcing, and investigates to what degree OEMs buy production equipment in the UK, so called 'non-production' or 'secondary' purchasing.

The typical spend of UK vehicle manufacturers on machinery, robots, and assembly equipment is £17m per annum. This figure is **two orders** of magnitude smaller than the average volume spent on component purchasing, in that sense a comparatively small opportunity for the UK economy, accordingly this section of the report will be brief.

Spend on machinery, robots and assembly equipment in £m	Min	Max	Mean	Total	N
	£2m	£80m	£17m	£104m	6

The degree to which production equipment is bought from the UK is rather low, both in absolute and relative terms. When asked whether any of the following were bought from UK companies, the OEMs response was as follows:

Answer	N	%
Machinery and equipment	7	78%
Assembly lines / equipment	6	67%
Tooling	6	67%
Robots	4	44%

The key reasons why such equipment is not bought in the UK are fourfold: an offer that is not competitive in price, in specification, no suppliers available in the UK, and that equipment suppliers are prescribed by headquarters.

SUMMARY OF KEY FINDINGS

The key findings of the OEM survey are:

1. There is both scope and strategic intent to increase UK sourcing. The actual responsibility for sourcing decisions in the UK follows a bimodal distribution, however all OEMs claim that their UK operations have some say in making this decision.

2. There exists an extensive OEM wish list of items that OEMs would like to procure in the UK, in addition to their current sourcing (see Section IV for details).

3. The main reason why OEM purchase in the UK is proximity, which accounts for 69% of total responses: this includes lower logistics cost, the configuration of parts, and the support of UK-built vehicles.

Several questions have remained unanswered from this survey, and these will be taken forward in the survey of component suppliers, reported on next:

4. The main reason for not sourcing from the UK is that unit cost is seen as uncompetitive by the OEM. This finding as such does not yield any new insights, and needs to be further clarified in the component supplier survey.

5. The OEMs clearly stated why they source from the UK, yet it is not clear what the component suppliers consider as their competitive advantage. It is vital to understand whether there is indeed congruence in views as to what the OEM consider as the competitive edge of the UK supply chain, and what component suppliers regard as such.

6. The OEM research, by default, has focused on the link to their Tier-1 suppliers only, yet we also need to understand what the constraints are in tiers 2 and 3 of the supply chain.

7. Furthermore the OEM survey highlighted questions as to which level of value-added is actually conducted by UK suppliers. In other words, it is possible to 'buy' components assembled or configured in the UK, while the actual manufacturing of these may have taken place outside the UK.

8. Finally, there is a general need to triangulate, and validate, the OEM findings with the results from the supplier survey.

IV. A SURVEY OF UK AUTOMOTIVE SUPPLIERS

Survey design

The key methodological problem when surveying automotive suppliers is to define a representative and clearly defined subset of firms to survey, as it was not feasible to identify and approach all 3,000 UK firms that are classified as 'automotive' in statistical terms. Instead, we used the supplier database kindly provided by the Society for Motor Manufacturers and Traders, which holds 6,200 component suppliers that are all (1) certified, (2) classify themselves as automotive, and (3), have a UK base. In total 620 suppliers were approached, of which several contact details were no longer available. Following the initial approach, and two rounds of follow-up, in total N=140 suppliers responded, which represents a 22.5% response rate.

In terms of distribution by tier, 42% classify themselves as first tier supplier, while 19% as second-tier supplier, and 34% as both first- and second-tier supplier.

Classification		Response	%
First-tier supplier (mostly directly supplying vehicle manufacturers)?		59	42%
Second-tier supplier (mostly supplying first-tier suppliers)?		26	19%
Both first and second-tier supplier?		47	34%
Raw materials supplier (including processing)?		5	4%
Other		3	2%
Total		140	100%

In terms of distribution across the seven component categories discussed earlier, the supplier survey achieved a fairly even spread, which allows us to comment on all component categories, see table below.

	Response	%
Powertrain	54	39%
Chassis	54	39%
Interior	61	44%
Exterior	47	34%
Electrical	20	14%
Electronics	17	12%
Consumables	8	6%
Other	1	1%

Level of Value-added

A key consideration for this study is the level of value created in the automotive supply chain, as there is less economic benefit in developing more 'screw-driver' factories (assembly and configuration sites), as opposed to true manufacturing operations.

Overall we find that 74% of suppliers manufacture in the UK. At first tier level this level is lower at 65%, while at second tier virtually all (97%) of suppliers also manufacture. The table below shows the profile of value-adding activities, in total and by supplier tier.

	Manufacturing (component)	Assembly	Late configuration/ Sequencing (e.g. supplier park)	Research and development	Warehousing and distribution	Sales office
All Respondents (in % total responses)	74%	61%	21%	38%	42%	51%
Tier 1 (in % total responses)	31%	33%	14%	16%	17%	20%
Tier 1 (in % total Tier I responses)	65%	69%	28%	32%	36%	41%
Tier 2 (in % total responses)	13%	8%	1%	5%	6%	7%
Tier 2 (in % total Tier II responses)	97%	61%	9%	33%	42%	48%

Note: Data includes information of multiple plants per individual respondent

In terms of number of plants, the majority (56%) of suppliers operate one plant only in the UK, while - as one would expect - first-tier suppliers are generally larger in size, and thus operate comparatively more plants.

Number of plants	1	2	3	4	5	6	Total responses
All	139	49	23	17	11	5	244
Tier 1	59	21	14	11	8	4	117
Tier 2	26	5	1	1	0	0	33

The importance of the automotive business

Many suppliers (in particular further upstream) work across industry sectors, it is thus important to assess the general importance of the automotive business for the UK supply base, as a fraction of their overall business. The average fraction of automotive turnover is very high, at 78%, but can be as low as 3%. The average for first-tier suppliers is 88%, compared to 68% for second-tier suppliers.

UK turnover by automotive customers (in %)	Min	Max	Mean	Std Dev.	N
All	3	100	78	31	139
Tier 1	3	100	88	25	59
Tier 2	5	100	68	30	26

CURRENT SUPPLY

In this section we will review the current supply patterns of UK suppliers, before turning to their competitiveness, business development aspects, and the sourcing suppliers conduct in the UK.

Main customers

The number of automotive customers served from the UK plants differs considerably between first- and second-tier suppliers. In this part of the survey we did find a great deal of variation in responses, as shown below, hence will comment on the median results only. Here, the average supplier serves 6 automotive customers, which does not significantly differ across tiers (see table below for details).

It should be noted that although the minimum number of automotive customers for most categories is 'zero', this does not mean that these supplier do not undertake automotive business at all. At the least, all suppliers surveyed one or more automotive customers. However, some suppliers only do server UK-based customers, while others exclusively serve customers abroad.

Number of customers	Min	Max	Mean	Median	Std Dev.
ALL					
Number of automotive customers in the UK	0	300	14	6	31
Number of automotive customers abroad	0	500	18	5	58
TIER 1 ONLY					
Number of automotive customers in the UK	0	117	7	5	15
Number of automotive customers abroad	0	275	11	4	37
TIER 2 ONLY					
Number of automotive customers in the UK	1	46	10	7	10
Number of automotive customers abroad	0	60	9	4	16

As one would expect, the UK OEMs feature very strongly on the list of main customers for UK suppliers. In terms of percentage of automotive business, the table below shows the importance of the respective OEMs to the UK supply base.

Main customer	Percentage of turnover		
	Mean	Std Dev	N
BMW-Mini	25	29	22
Jaguar Land Rover	22	19	53
Toyota	22	23	38
Nissan	20	17	29
Honda	20	21	27
Ford	18	16	31
Volvo	15	17	14
Leyland Trucks	14	8	2
Renault	11	10	10
GM	12	12	12
JCB	10	5	4
Bentley	7	9	11
PSA	5	3	5
VW	6	7	7
Aston Martin	6	8	8

COMPETITIVENESS

In this section we asked UK suppliers to specify why they manufacture in the UK, in order to distil the key competitive advantage for UK-based manufacturing. As can be seen, the notion of 'proximity' again features very strongly in the responses, thus mirroring the findings from the OEM survey. Other features include labour (25%), unique processing capabilities (15%), and high quality (11%), as well as other (10%, see list below). In comparison though proximity related factors account for 40% overall.

Answer	Response	%
To support our customers in the UK	82	60%
We are a UK-based company	71	52%
Legacy / historic reasons	65	47%
Logistics requirements	45	33%
Availability of qualified labour	25	18%
The vehicles we supply to are only made in the UK	22	16%
Unique processing capabilities in the UK	15	11%
The UK is the highest quality location	11	8%
Other (see below)	10	7%
Availability of qualified suppliers	9	7%
The UK is the lowest cost location	9	7%
Proximity to energy sources	4	3%
Proximity to raw material sources	3	2%

When asked for their specific, perceived main competitive advantage, suppliers quoted the following (multiple responses were possible):

	Response	%
Our quality	113	82%
Our delivery reliability and responsiveness	108	78%
Our manufacturing cost	72	52%
Our R&D capabilities	54	39%
Our low logistics / delivery cost	52	38%
Our unique technology / patents	38	28%
Our products need to be configured close to the vehicle assembly plant	33	24%
Other (see below)	12	9%

Other multiple mentions included the local infrastructure, knowledge and skills, being a full-service provider, and reputation.

What transpires from the above table are the standard elements of performance, QCD (quality, cost, delivery), while again proximity as well as knowledge and skills feature very strongly.

OEM supplier selection

In terms of how suppliers are being evaluated by their OEM customers, the following criteria are being used. Again, QCD (not unexpectedly) features very strongly. In addition, innovation, currency stability, labour cost inflation are mentioned. Further mentions include the skill base, the ability to meet the annual cost reduction targets.

	N	%
Quality performance	130	94%
Reliability of supply / delivery performance	116	84%
Logistics cost to customer site	68	49%
Innovation	64	46%
Labour cost	38	28%
Currency stability	36	26%
Other (see below)	11	8%
Labour cost inflation	7	5%

Why business is lost

When a competitive tender is lost, the main reasons the customers provide tends to be a unit cost that was not competitive. Again, this answer mirrors the OEM survey results, and is of little use for policymaking as it leaves no real 'levers' to change this situation.

	N	%
Unit cost was not competitive	119	89%
Other (see below)	19	14%
Delivery / logistics was not competitive	14	11%
Volumes / capacity was insufficient	9	7%
Finance not available	8	6%
Required materials processing capabilities not available	7	5%
Required raw materials not available	3	2%
Quality was not competitive	1	1%

It is thus interesting to note that the reasons for lost business (culminating in the overall unit cost) are multiple, and varied in nature: from poor logistics and quality, to insufficient capacity and finance, and processing capabilities that were not available. Further issues mentioned on multiple occasions include: OEM supply strategy, tooling investment that was too high, OEMs staying with the incumbent supplier, concerns over small business size, and financial stability aspects.

Thus we can see that (1), operational (QCD) and (2), financial aspects stand out as key reasons why UK suppliers lose business.

Where is business lost to?

When automotive business is lost, it tends to evenly go to three regions: (1) Western Europe, including the UK (35%), low-cost countries in Eastern Europe and BRIC countries (36%), while the remainder follows no clear pattern. This overall is good news, as it means that labour cost is not the only determinant for sourcing decisions, which in turns means that at least one third of this business could potentially be won back by UK suppliers.

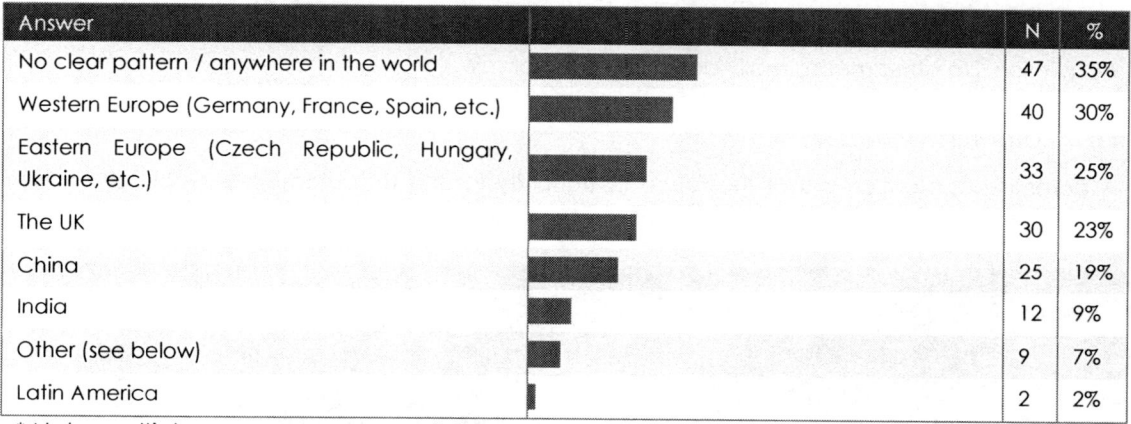

Answer	N	%
No clear pattern / anywhere in the world	47	35%
Western Europe (Germany, France, Spain, etc.)	40	30%
Eastern Europe (Czech Republic, Hungary, Ukraine, etc.)	33	25%
The UK	30	23%
China	25	19%
India	12	9%
Other (see below)	9	7%
Latin America	2	2%

* Note: multiple responses were possible

SUPPLIER BUSINESS DEVELOPMENT

In this section we assess the strategic growth areas for suppliers, where we aim to identify areas of overlap between OEM needs, and supplier growth areas. This data was also core to the development of the sourcing roadmap shown in Section VI, which matches OEM needs with supplier intentions.

We furthermore investigated how suppliers approach OEMs, and found that both a proactive approach, as well as participating in a bidding process, are the two main means for business acquisition. No other means really does play a major role.

Low carbon powertrain parts

A common contention is that the future opportunity for the UK automotive supply chain lies with the development and large-scale production of low-carbon vehicles.

Question	Strongly Disagree	Slightly Disagree	No opinion either way	Somewhat Agree	Strongly Agree	Mean	Std Dev.	CI[1] (±)
We are aware that our main OEM customers are developing low-carbon powertrains	3	2	21	37	68	1.25	0.94	0.16
We have received considerable interest from our OEM customers with regard to supplying their low carbon vehicles	20	20	27	33	29	0.24	1.37	0.24
Within five years time, a significant proportion of our production will be geared to support low-carbon vehicle production	15	26	31	40	17	0.12	1.25	0.22
Within our company, our UK operations is at the forefront of developing low-carbon technologies	40	28	30	19	13	-0.48	1.33	0.23

[1] – 95% confidence interval

As the table shows, however, UK suppliers are currently not well placed to harness this opportunity: while UK suppliers are aware of the developments their OEM customers are undertaking, they are neither asked to participate, nor do they feel that within their own firm they are at the forefront of this development.

While this finding is discouraging in some aspects, it should not be forgotten that we are still far away from a volume-based production of low-carbon powertrains, and as such, this development can be still be reversed, in our view.

CURRENT SUPPLIER SOURCING

In this section we will assess the sourcing patterns of UK suppliers, in other words, what UK suppliers buy in the UK and abroad. A secondary aim is to identify supply chain constraints that may exist further upstream, i.e. a 2nd or 3rd tier level in the supply chain.

Current supplier sourcing in the UK

In terms of average purchasing spend, the UK supplier buys on average £36m of 26 suppliers, of which 46% is spent in the UK.

	Min	Max	Mean	Median	Std Dev.
Approximate purchasing budget per annum for UK operations	0	500	36	8	86
Percentage of this volume sourced in the UK	0	100	46	40	32
Number of major production suppliers	0	190	26	12	34

In terms of outlook, the following table shows that UK suppliers are not foreseeing any major change in the sourcing patterns from their UK suppliers. It is also not their strategic intent to increase their local sourcing, in contrast to what the OEMs responded. They feel well in charge to make their own sourcing decisions, and are less bound by central or group considerations, in comparison to the OEMs.

Question	Strongly Disagree	Slightly Disagree	No opinion either way	Somewhat Agree	Strongly Agree	N	Mean	Std Dev.	CI[1] (±)
The volume we purchase from our UK suppliers is likely to increase in the next five years	13	25	38	41	7	124	0.03	1.09	0.19
It is our strategic intent to increase our automotive business in the UK	5	12	21	32	52	122	0.93	1.17	0.21
We have little influence over supplier selection as this is done outside of the UK	43	23	14	31	10	121	-0.48	1.41	0.25
When we put out a call for tender, we receive competitive bids from UK suppliers	5	32	41	35	8	121	0.07	0.99	0.18
Our UK operation is for manufacturing only, all major strategic decisions are taken at headquarters	67	14	14	18	5	118	-1.02	1.30	0.23

[1] – 95% confidence interval

Key supply chain constraints

In terms of key constraints in the UK supply chain, suppliers point to two main factors: an overall unit cost that is not competitive, and a general lack of accredited suppliers. Interestingly, these constraints do mirror the points made by OEMs also. In particular, short supply are raw materials, and manufacturing equipment suppliers.

Components	Unit cost not competitive	Quality not competitive	Delivery / logistics not competitive	Volumes / supplier capacity insufficient	Lack of accredited suppliers in the UK	Required materials processing capabilities not available
Components and parts	75	23	10	13	22	17
Raw materials	50	10	10	12	26	27
Consumables: sealers, grease, adhesive, lubricants	20	4	3	3	7	6
Hardware: nuts, bolts, screws, fasteners	38	4	5	3	7	3
Tooling	70	9	8	6	19	7
Manufacturing equipment	46	11	5	6	25	14
Total	299	61	41	43	106	74

SUMMARY OF KEY FINDINGS

The key findings of the supplier survey are:

1. The findings of the supplier survey largely mirror the earlier findings of the OEM survey, in as far as the same supply chain constraints that exist at first tier level, also permeate the higher levels of the supply chain. For example, basic materials processing capabilities in terms of pressing, forging and casting are mentioned by both suppliers and OEMs as constraints.

2. On the positive side, UK suppliers do see a wide range of potential growth areas, which in many ways overlap with the OEM needs identified earlier. This in turn provides a great opportunity for 'match making', which will be taken up further in the sourcing roadmap.

3. The reasons why supply business is lost in the UK could be further illuminated in the supplier survey, however, it is not possible to identify any uniform pattern here. The main reasons however relate to operational execution (QCD), as well as financial aspects (availability of finance, concerns over supplier stability)

4. What is known, however, is that only one third of supply business is lost to low-cost countries, while one third stays within Western Europe, which does provide an opportunity to win significant parts of this business back to the UK.

V. THE SOURCING ROADMAP

It is difficult to devise industrial policy interventions that specifically support the entire component supply chain of an industrial sector: General 'support' programmes are often too generic, and are hard to assess in terms of their actual impact. Supporting specific ventures on the other hand is problematic from a state aids and competition regulation point of view. This 'sourcing roadmap' in turn aims to bridge this gap, by providing empirical evidence of the perceived competitiveness and constraints that are seen by both manufacturers, and suppliers. By triangulating their views, this report not only reduces the likely bias any individual firm's response would induce, it also distils common patterns across tiers, and component groups.

This report does not make any recommendations as such, it points towards the key opportunities for retaining, and building, supply chain capabilities. The Automotive Council, and its Supply Chain and Technology subgroups, will devise specific recommendations on the basis of this report, and other evidence.

Summary of Combined Findings across both OEM and Supplier Surveys

The UK automotive supply chain largely supports the vehicle programmes assembled in the UK, as one would expect, yet it also exports considerable amounts. At present, about 80% of all component types required for vehicle assembly operations can be procured from UK suppliers.

In terms of size, the combined UK purchasing spend of the UK-based automotive, commercial vehicle and yellow goods manufacturers is £7.4bn per annum. This equates to about 36% of their global purchasing spend.

In terms of value-added at supplier level, on average 74% of UK-based suppliers manufacture (as opposed to assemble, late configure, or distribute components) in the UK; this ratio is lower at first tier suppliers where 65% manufacture, while virtually all second-tier suppliers operate manufacturing facilities in the UK. The average supplier serves 6 customers (median), with a strong towards bias towards those OEMs that operate vehicle and engine assembly plants in the UK.

In terms of competitiveness, the notion of 'proximity' was identified as the key competitive advantage of UK suppliers: in operational terms, proximity allows for (1) lower logistics cost, a better support for UK-built vehicles, (2) the responsive configuration of parts, as well as (3) for more flexibility to adjust to volume and product mix fluctuations. In strategic terms proximity also acts as (4), a general proxy for risk reduction the supply chain, as well as (5), a hedge against currency fluctuations.

The key reason why UK suppliers are losing business is that their unit cost was not competitive. The OEMs further state that secondary reasons for not sourcing from UK suppliers were, in order of prevalence, (a) the general lack of accredited suppliers, (b) required processing capabilities that were not available, (c) quality and (d) logistics that were not competitive. While it is not possible to identify any specific patterns here, the main reasons however relate to operational execution (QCD), as well as financial aspects (availability of finance, concerns over supplier stability).

When UK supply business is lost, about one third stays within the UK and Western Europe, one third goes to low-cost countries, while the final third show no clear pattern. The risk of losing business to low-cost regions rises however considerably for second-tier suppliers. Nonetheless, overall this does provide the opportunity to win significant fractions of this business back to the UK.

The opportunities

The combined manufacturer and suppliers surveys have identified three 'clusters' of opportunities for building, and retaining supply chain capabilities in the UK. These three clusters are:

1. 'Classic' components (such as trim, mouldings, struts, glass, etc) where a clear match between OEM requirements and supplier growth areas could be identified.

2. 'Electric powertrain' components (such as batteries, motors, inverters, etc.) which will be increasingly required to the low-carbon vehicle production.

3. 'Heavy metal' components, (such as castings, forgings, pressings, wheels, bearings, etc) which were mentioned by both OEMs and suppliers as key constraints in their respective supply chains.

While these are all automotive 'parts', these are very different clusters in nature, and addressing these opportunities will also have to take place on different time scales. In short, each requires a specific strategy.

There is great potential for the 'classic parts' cluster, by assembling the matched needs of OEMs and supplier growth areas, and to foster interaction between these parties. Not unlike a 'match making' arrangement, a key objective here is to bring UK suppliers together with OEM purchasing directors. Given that the components in question are all mature technologies, and are demanded for the (ongoing) volume production, such activities could start immediately.

Matters are different for the 'electric powertrain' cluster, which is supporting the shift towards (full or partly) electric powertrain configurations. This technology is still comparatively new, and as such, demand for these components is still low, albeit growing. The challenge in this cluster is to establish a supplier base in the UK that will be able to meet this growing demand in the future. Key components include batteries, hub motors, power electrics, converters, and the like. A main approach could be to initially identify the 'Top-10' firms globally, and approach these individually to spark their interest in establishing a production base in the UK.

The third cluster, 'heavy metal' components, in many ways provides the greatest challenge, as the manufacturing sectors in question are scale dependent industries, with a low innovation clock-speed. Interviews with sector experts have reaffirmed that in the past many of the UK suppliers in this area have closed or relocated their operations to low-cost countries, primarily because of lower energy cost and more lenient regulation in terms of emissions, as well as health and safety. The economic viability of bringing this capability back the UK thus seems rather questionable.

The 'UK Sourcing Roadmap'

In order to summarise the three opportunities outlined above, and provide a more concise framework for addressing these, a 'UK Sourcing Roadmap, shown below, has been assembled, which visualises the key short-term and medium-term opportunities, as well as the areas where critical support is required to the support the UK automotive supply chain.

Short term opportunities largely arise where OEM sourcing needs match with strategic growth areas of UK suppliers, in other words, where OEMs have a current need that could potentially be met from a UK-based supplier. Here, considerable potential was identified for the 'classic' components sourced. A comprehensive list of commodities that OEMs wish to procure was developed as a guide to identify specific supply opportunities (See Appendix A).

Medium- to long term opportunities for building the UK supply chain arise from the gradual shift towards a portfolio of powertrain architectures, and the novel components these will require. At present, UK suppliers do not feel they are at the forefront of supporting this shift. Here, the 'Top-10' most desirable low carbon powertrain suppliers will be identified, in order to target efforts to entice these to commence operation in the UK.

Finally, in terms of further critical support areas, the surveys show that UK suppliers are losing out on a unit cost basis, while – paradoxically – proximity was rated as the most important competitive factor. Hence a key opportunity arises to help suppliers conceptually to make their business case on a 'total supply chain cost' basis, rather than rely on unit cost calculations alone, which will largely ignore their key competitive advantage in competing for business.

Outlook

Devising industrial policy interventions to support any critical industry is not a trivial task, as firm-level idiosyncrasies make it hard to distil the underlying patterns and problems that need to be addressed in order to retain, and build, supply chain capabilities in the UK.

By analysing the current sourcing patterns of both vehicle manufacturers and suppliers, by understanding their respective assessments of perceived supplier competitiveness and wider supply chain constraints, this report provides the empirical grounding for targeted policy interventions that will contribute towards retaining a healthy automotive industry in the UK.

UK Sourcing Roadmap

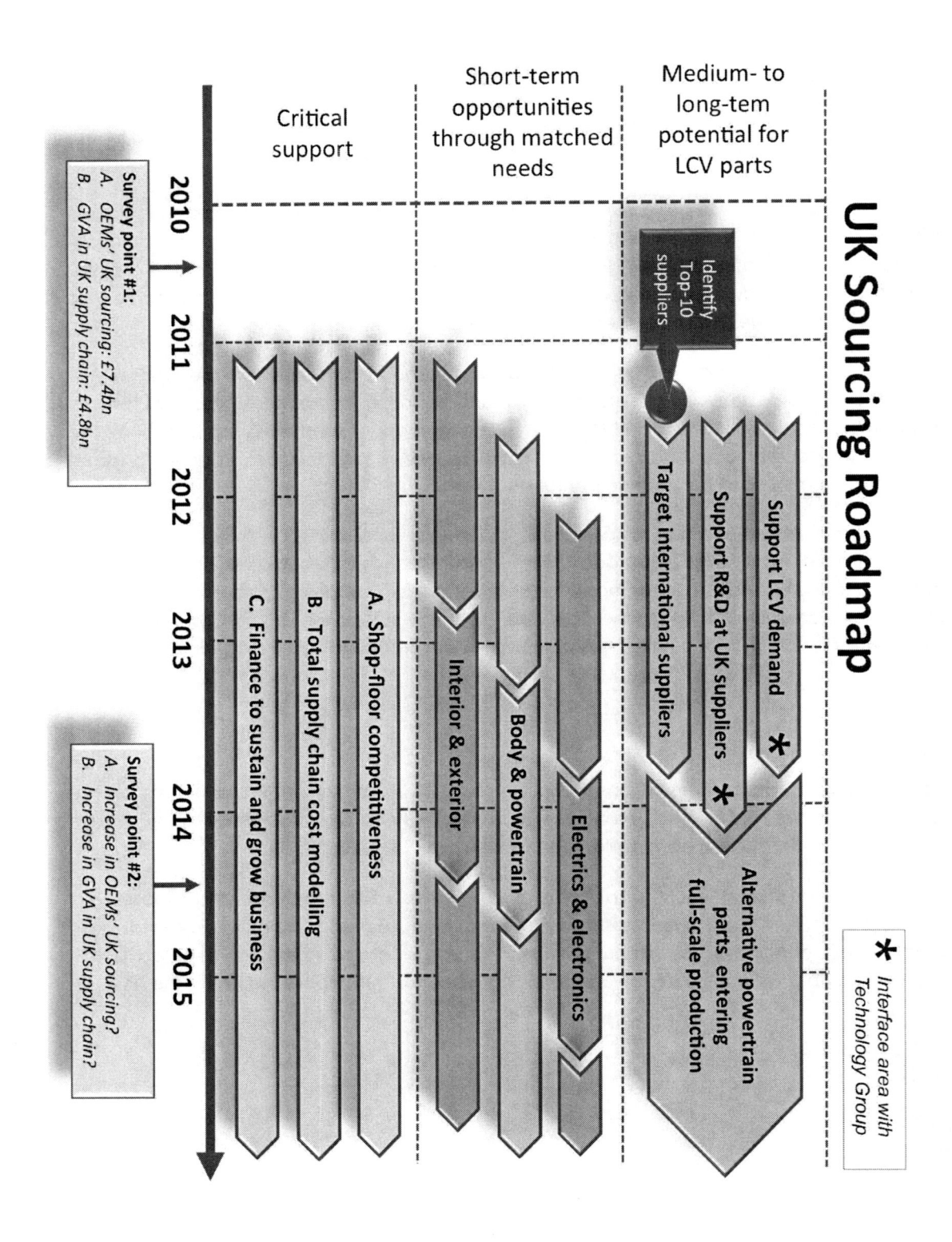

Critical support
- A. Shop-floor competitiveness
- B. Total supply chain cost modelling
- C. Finance to sustain and grow business

Short-term opportunities through matched needs
- Interior & exterior
- Body & powertrain
- Electrics & electronics

Medium- to long-tem potential for LCV parts
- Support LCV demand *
- Support R&D at UK suppliers *
- Target international suppliers
- Identify Top-10 suppliers
- Alternative powertrain parts entering full-scale production

Timeline: 2010, 2011, 2012, 2013, 2014, 2015

Survey point #1:
A. OEMs' UK sourcing: £7.4bn
B. GVA in UK supply chain: £4.8bn

Survey point #2:
A. Increase in OEMs' UK sourcing?
B. Increase in GVA in UK supply chain?

* Interface area with Technology Group

APPENDIX A: OEM SOURCING REQUIREMENTS

This appendix lists the requirements that OEMs have communicated to us in the course of this study. The first part (A1) lists all parts that OEMs would like to source from the UK in the future, the second part (A2) specifically lists those components for which, according to OEM perception, there currently are no UK sources.

Appendix A1: OEM Sourcing Requirements, by component category, with combined annual volumes [in number of units].

1. Body & Powertrain

1.1 Powertrain

Castings: Aluminium and Iron	550,000
Chain case	250,000
Crankshafts	20,000
Flywheel	10,000
Engine commodities	volume tbc
Gear boxes	volume tbc
Oil pans	250,000
Heavy metal: forgings, large castings	1,000,000

1.2 Body

Aluminium pressing & assembly	100,000
Badges	volume tbc
Exhaust hangers	966,000
Fuel filler cap (petrol & diesel)	322,000
Heat shields (Fuel tank & exhaust)	452,000
Hinges	1,600,000
Hot stampings	volume tbc
Large stamping	476,000
Pressed metal structures	2,000
Latches	800,000

1.3 Chassis, braking, steering & suspension

Alloy Wheels / Finish Wheels	1,400,000
Wheels	5,000,000
Tyres	volume tbc
Wheel bearings	800,000
Brake systems and components (eg cables, discs, tube, pedals)	volume tbc
Parking brake device Standard & mechanical	322,000
Clutch pedal Assembly (non plastic, non modular)	322,000
Drive shafts	560,000
Power springs	500,000
Suspension springs	1,000,000
Shock absorbers (Struts)	5,300,000
Steering systems (Steering gears, columns etc)	500,000
Corner unit module	volume tbc
General tubular assemblies	250,000

2. Interior & Exterior

2.1 Interior

Carpets	volume tbc
Clip (trim)	1,000,000
Headliners, Fabric and Foam (Headliners etc)	450,000
Seat components (L2 parts fabric, foam pads, headrests etc)	250,000
Instrument panels	100,000
Interior trims (door, sun visors etc)	7,500,000
Door trim	100,000
Internal mirrors	250,000
Appliance trims	30,000
Architectural trims	5,000
Blow mouldings	250,000
Chroming (plastic)	volume tbc
Med/Large injection mouldings	250,000

2.2 Exterior

Large external mouldings (mirrors, bumpers etc)	500,000
Bumper parts	1,000,000
Glass	7,000,000

3. Electrical & Electronics

3.1 Electrical

Heat and air conditioning units	volume tbc
Alternators	250,000
Batteries 12V 24V	volume tbc
Harness connectors	250,000
Starter motors	250,000
Wiper Systems	volume tbc

3.2 Electronics

Electronic control units	volume tbc

4. Low Carbon/Hybrid/EV	
Batteries - EV	269,000
Fuel Cells	100,000
Hybrid Conversion	200
Hybrid Fuel Systems	5,000
Range extender engines	100,000
Reducers	100,000
Inverters	100,000
Large electric motors	100,000
Charging Technology - EV	19,000
Parking brake system EV	50,000

5. Additional components	
Bearings	7,000,000
Bolts	1,000,000
Non locking Wheel Nut	2,500,000
Nuts (Fasteners)	1,000,000
Machining	volume tbc
Tool making (plastic injection, stamping etc)	volume tbc
Sheet steel	volume tbc
Air cleaners	250,000

Appendix A2: OEM Sourcing Requirements, by component category, for which - according to the OEMs' perception - UK sources currently are not available.

This second part provides a list of components like to source from the UK in the future, provided a competitive supplier was available.

1. Body & Powertrain

Component	Annual volume [units]
Forgings	250,000
Castings	250,000
Bearings	5,000,000
Wheels	5,800,000
Chassis/Shock absorbers	5,250,000
Dampers	250,000

2. Interior & Exterior

Component	Annual volume [units]
Interior trim sets	7,400,000
Sunvisors	2,400,000
Sub frames	12,000
Mirrors	250,000
Ext. mouldings (all sizes: mirrors to bumpers)	500,000
Glass	7,006,000
Bumpers	6,000
Soft tops	50,000
Hot stampings	200,000
Exterior painted mouldings (mirror caps, spoilers etc)	250,000

3. Electrics & Electronics

Component	Annual volume [units]
Electric motors	100,000
Batteries	56,000
Large electric motors	100,000
Harnesses	6,000

4. Low Carbon/Hybrid/EV

Component	Annual volume [units]
Charging technologies	50,000
Reducese and invertors	50,000

APPENDIX B: OEM SOURCING PATTERNS, BY COMPONENT CATEGORY

How to read the data?

The data in this section provides an in-depth overview of the sourcing patterns of UK-based vehicle manufacturers, by component category. The data is firm-level data, and is not volume-corrected, as the main focus on is firm-level decision-making, not on the overall sourcing volumes from UK component suppliers.

The first table is a summary for this component group (of which there are seven), and shows indices (with values of 0=no, and 1=yes) for:

1. Whether the part is currently sourced in the UK
2. This source is a single source
3. This part is sourced from an internal supplier (with "partly" giving a value of 0.5)

The next column shows the mean annual volume in units, while the final column shows the perception of whether the volume for this component is expected to change, and if so, in what direction (with values of -2 ("strong decline") to 2 "strong growth", and 0 being "no change").

The following three tables (#2-#4) shows the same data as in the first table, yet at greater detail for each component in that component group.

Powertrain: Current sourcing

Components	UK sourced	Single source	in-house supplier	Approx. annual volumes	Expected volume change -/+
	Mean	Mean	Mean	Mean	Mean
Engine components	0.80	0.30	0.00	284,500	0.55
Transmission components	0.20	0.30	0.13	294,500	0.45
Forgings	0.22	0.33	0.00	1,104,600	0.60
Castings	0.56	0.22	0.11	1,504,600	0.50

These parts are currently sourced from the UK

Components	Yes	No	Mean
Engine components	8	2	0.80
Transmission components	2	8	0.20
Forgings	2	7	0.22
Castings	5	4	0.56

Approx. annual volumes

Components	Total	Min	Max	Mean
Engine components	1,707,000	5,000	1,200,000	284,500
Transmission components	1,767,000	5,000	1,200,000	294,500
Forgings	5,523,000	5,000	5,000,000	1,104,600
Castings	7,523,000	5,000	7,000,000	1,504,600

The volume for this part is likely to

Components	Decrease	Remain Stable	Increase	Mean
Engine components	0	5	6	0.55
Transmission components	1	4	6	0.45
Forgings	0	4	6	0.60
Castings	1	3	6	0.50

Exterior: Current sourcing

Components	UK sourced	Single source	in-house supplier	Approx. annual volumes	Expected volume change -/+
	Mean	Mean	Mean	Mean	Mean
Panels	0.60	0.22	0.29	898,714	0.56
Bumpers	0.60	0.56	0.30	4,046,167	0.56
Spoilers & body cladding	0.70	0.30	0.00	117,000	0.50
Glass	0.50	0.20	0.00	1,227,833	0.40

These parts are currently sourced from the UK

Components	Yes	No	Mean
Panels	6	4	0.60
Bumpers	6	4	0.60
Spoilers and body cladding	7	3	0.70
Glass	5	5	0.50

Approx. annual volumes

Components	Total	Min	Max	Mean
Panels	6,291,000	5,000	5,000,000	898,714
Bumpers	24,277,000	5,000	24,000,000	4,046,167
Spoilers and body cladding	702,000	1,000	250,000	117,000
Glass	7,367,000	5,000	7,000,000	1,227,833

The volume for this part is likely to

Components	Decrease	Remain Stable	Increase	Mean
Panels	0	4	5	0.56
Bumpers	0	4	5	0.56
Spoilers and body cladding	0	5	5	0.50
Glass	1	4	5	0.40

Interior: Current sourcing

Components	UK sourced	Single source	in-house supplier	Approx. annual volumes	Expected volume change -/+
	Mean	Mean	Mean	Mean	Mean
Seats	0.64	0.36	0.13	918,833	0.45
Instrument panels	0.55	0.50	0.22	248,500	0.50
HVAC units	0.45	0.55	0.09	248,000	0.45
Headliners	0.73	0.36	0.00	248,500	0.45
Carpets	0.64	0.45	0.00	3,381,833	0.45
Interior trim	0.73	0.09	0.00	1,718,833	0.45

These parts are currently sourced from the UK

Components	Yes	No	Mean
Seats	7	4	0.64
Instrument panels	6	5	0.55
HVAC units	5	6	0.45
Headliners	8	3	0.73
Carpets	7	4	0.64
Interior trim	8	3	0.73

Approx. annual volumes

Components	Total	Min	Max	Mean
Seats	5,513,000	5,000	5,000,000	918,833
Instrument panels	1,491,000	5,000	1,200,000	248,500
HVAC units	1,488,000	3,000	1,200,000	248,000
Headliners	1,491,000	5,000	1,200,000	248,500
Carpets	20,291,000	5,000	20,000,000	3,381,833
Interior trim	10,313,000	5,000	10,000,000	1,718,833

The volume for this part is likely to

Components	Decrease	Remain Stable	Increase	Mean
Seats	0	6	5	0.45
Instrument panels	0	5	5	0.50
HVAC units	0	6	5	0.45
Headliners	0	6	5	0.45
Carpets	0	6	5	0.45
Interior trim	0	6	5	0.45

Chassis: Current sourcing

Components	UK sourced	Single source	in-house supplier	Approx. annual volumes	Expected volume change -/+
	Mean	Mean	Mean	Mean	Mean
Brakes (discs, drums)	0.00	0.27	0.00	886,833	0.45
Suspensions (struts)	0.27	0.36	0.10	881,167	0.45
Fuel tanks	0.91	0.45	0.09	248,667	0.45
Wheels	0.00	0.36	0.00	895,500	0.45
Tyres	0.18	0.00	0.00	895,500	0.45

These parts are currently sourced from the UK

Components	Yes	No	Mean
Brakes (discs, drums)	0	11	0.00
Suspensions (struts)	3	8	0.27
Fuel tanks	10	1	0.91
Wheels	0	11	0.00
Tyres	2	9	0.18

Approx. annual volumes

Components	Total	Min	Max	Mean
Brakes (discs, drums)	5,321,000	5,000	5,000,000	886,833
Suspensions (struts)	5,287,000	2,000	5,000,000	881,167
Fuel tanks	1,492,000	5,000	1,200,000	248,667
Wheels	5,373,000	5,000	5,000,000	895,500
Tyres	5,373,000	5,000	5,000,000	895,500

The volume for this part is likely to

Components	Decrease	Remain Stable	Increase	Mean
Brakes (discs, drums)	0	6	5	0.45
Suspensions (struts)	0	6	5	0.45
Fuel tanks	0	6	5	0.45
Wheels	0	6	5	0.45
Tyres	0	6	5	0.45

Electronics: Current sourcing

Components	UK sourced	Single source	in-house supplier	Approx. annual volumes	Expected volume change -/+
	Mean	Mean	Mean	Mean	Mean
Anti-lock brakes (ABS)	0.00	0.40	0.00	877,833	0.40
Engine control unit (ECU)	0.20	0.50	0.11	248,500	0.50
Entertainment / radio systems	0.00	0.33	0.00	247,167	0.50

These parts are currently sourced from the UK

Components	Yes	No	Mean
Anti-lock brakes (ABS)	0	10	0.00
Engine control unit (ECU)	2	8	0.20
Entertainment / radio systems	0	10	0.00

Approx. annual volumes

Components	Total	Min	Max	Mean
Anti-lock brakes (ABS)	5,267,000	-	5,000,000	877,833
Engine control unit (ECU)	1,491,000	5,000	1,200,000	248,500
Entertainment / radio systems	1,483,000	2,000	1,200,000	247,167

The volume for this part is likely to

Components	Decrease	Remain Stable	Increase	Mean
Anti-lock brakes (ABS)	0	6	4	0.40
Engine control unit (ECU)	0	5	5	0.50
Entertainment / radio systems	0	5	5	0.50

Electrical: Current sourcing					
Components	UK sourced	Single source	in-house supplier	Approx. annual volumes	Expected volume change -/+
	Mean	Mean	Mean	Mean	Mean
Batteries	0.00	0.27	0.00	249,500	0.45
Alternators	0.09	0.45	0.00	248,500	0.45
Harnesses	0.00	0.36	0.00	1,734,833	0.45
Power steering	0.18	0.27	0.10	244,500	0.36

These parts are currently sourced from the UK

Components	Yes	No	Mean
Batteries	0	11	0.00
Alternators	1	10	0.09
Harnesses	0	11	0.00
Power steering	2	9	0.18

Approx. annual volumes

Components	Total	Min	Max	Mean
Batteries	1,497,000	5,000	1,200,000	249,500
Alternators	1,491,000	5,000	1,200,000	248,500
Harnesses	10,409,000	5,000	10,000,000	1,734,833
Power steering	1,467,000	-	1,200,000	244,500

The volume for this part is likely to

Components	Decrease	Remain Stable	Increase	Mean
Batteries	0	6	5	0.45
Alternators	0	6	5	0.45
Harnesses	0	6	5	0.45
Power steering	0	7	4	0.36

Consumables: Current sourcing

Components	UK sourced	Single source	in-house supplier	Approx. annual volumes	Expected volume change -/+
	Mean	Mean	Mean	Mean	Mean
Nuts, bolts, screws	0.50	0.20	0.00	20,252,200	0.50
Small plastics parts / fasteners	0.73	0.27	0.00	10,252,200	0.45
Adhesives and sealers	0.73	0.30	0.00	378,667	0.45

These parts are currently sourced from the UK

Components	Yes	No	Mean
Nuts, bolts, screws	5	5	0.50
Small plastics parts / fasteners	8	3	0.73
Adhesives and sealers	8	3	0.73

Approx. annual volumes

Components	Total	Min	Max	Mean
Nuts, bolts, screws	101,261,000	5,000	100,000,000	20,252,200
Small plastics parts / fasteners	51,261,000	5,000	50,000,000	10,252,200
Adhesives and sealers	2,272,000	1,000	2,000,000	378,667

The volume for this part is likely to

Components	Decrease	Remain Stable	Increase	Mean
Nuts, bolts, screws	0	5	5	0.50
Small plastics parts / fasteners	0	6	5	0.45
Adhesives and sealers	0	6	5	0.45

APPENDIX C: TOTAL SUPPLY CHAIN COST MODELLING

The benefits of global sourcing as part of a firm's purchasing strategy have been widely discussed: lower labour cost and efficient transportation systems generally promise great cost savings through sourcing from low-labour cost countries, or by offshoring factories into these regions. In practice however many firms fail to realise the cost advantages that global sourcing and offshoring ventures have promised. In this section we will explain the root causes underlying this phenomenon, and outline what opportunities arise for UK-based suppliers in explciting this when bidding for supply contracts to UK-based OEMs. Specifically, we will show that:

1. Global sourcing will (almost) always be cheaper in the short run, on the grounds of the labour cost differential

2. Dynamic supply chain costs, such as increased inventory levels, obsolescence and airfreight will lead to unexpected costs in global supply chains, but are unlikely to tip the balance in favour of domestic sourcing in the short run

3. Hidden cost, related to strategies risk and volatility in key business parameters such as currency exchange rates and labour cost inflation, are likely to tip the balance in favour of domestic sourcing in the medium- to long-term.

4. As a result UK suppliers will be at a perennial disadvantage when competing with global competitors on a unit cost basis, as their key competitive advantage – proximity – does not feature in the static costing models commonly used. Proximity and its related benefits only feature when total supply chain costs, as opposed to unit costs, are considered.

Over the past decades there has been an increasing trend towards offshoring manufacturing operations; typical destinations for relocation include BRIC (Brazil, Russia, India and China), and Eastern Europe ('East-shoring'). The primary motivation was a lower labour cost in these regions, coupled with reliable transportation links and reduced trade barriers. More recently however many firms have had to revisit their offshoring or global sourcing decisions, as the volatility in some parameters in their original cost calculations meant that many of the predicted savings did not materialise. Predominantly the volatility relates to labour cost inflation, the cost of transportation (driven by both fuel price volatility and container ship capacity constraints), and general uncertainty (traffic congestion, natural disasters, political unrest).

To illustrate, the figure below by McKinsey (2008) shows the average annual labour cost inflation in key manufacturing regions. As can be seen, annual labour cost increases of 20% are indeed likely in these 'manufacturing hotspots'.

A key reason why firms are now finding that their initial cost estimates of offshoring or global sourcing arrangements were not realistic was because the cost elements considered neglected some fundamental "dynamic" and "hidden" cost that do not factor in the most basic calculation, yet frequently occur in practice. We investigated this phenomenon through a set of exploratory cases (see Holweg et al., forthcoming), where we compare the planned and actual costs incurred cost five years after the decisions have been made.

A changing environment for offshoring

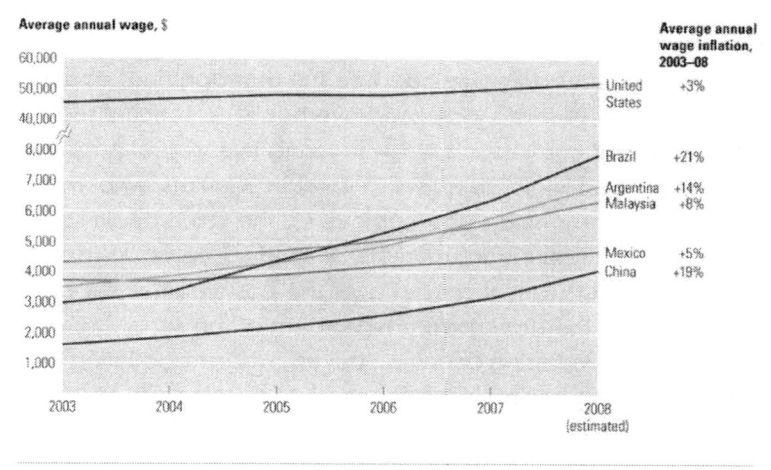

Average annual wage, $

Country	Average annual wage inflation, 2003–08
United States	+3%
Brazil	+21%
Argentina	+14%
Malaysia	+8%
Mexico	+5%
China	+19%

Source: Global Insight; Economist Intelligence Unit; CIBC World Markets; McKinsey analysis

Our findings showed that all firms had only considered "static costs", that is labour cost, materials and transportation costs, in their evaluation of their offshoring arrangements. However, virtually all of the unexpected costs resulted from dynamic distortions in the supply chain (with expedited shipments, or "airfreight", being the highest unexpected cost factor).

Such a *static* assessment omits the *dynamic* dimension of such sourcing decisions: the inherent assumption underlying an analysis that considers the *static* costs only is that demand is stable, and does not vary in the long-term. However, as customer demand invariably fluctuates to a certain degree, additional pipeline and safety stock will be required to counter this effect and ensure uninterrupted supply. Product variety amplifies the need for stockholding further. Also, the ability to introduce new products is seriously hampered as these additional stocks need to be cleared before new products can be introduced unless firms are willing to bear the resulting obsolescence. Other *dynamic* costs occur from stocks-outs and lost sales caused by long transport lead-times and the cost for obsolete materials ordered according to a long-term (yet commonly incorrect) forecast. Furthermore, the inability to customize products or build them to customer order might result in reduced margins, and/or higher rebates needed to sell these products in developed markets (Holweg and Pil, 2001, 2004). Finally, high cost for expedited shipments (such as air-freight) may be incurred when products are urgently required but are not shipped in time or their quality is found to be unacceptable.

A further cost category can be described as *hidden* costs; these are costs that are not related to the actual supply chain operation, but impact on the wider business environment, such as currency fluctuations, changing energy cost, and changes in the political climate or regulatory framework. Generally these costs are difficult to predict, and will incur on an irregular basis, such as costs resulting from currency fluctuations, additional travelling expenses required to coordinate the relationship or even the provision of resident engineers to solve ongoing problems. Furthermore, hidden costs are often not attributed to the individual ventures but are reflected in the general overheads, partly because the management accounting systems are not sophisticated enough to allocate increased overheads to specific supply relationships. In addition, there is a potential cost associated with losing intellectual property rights (IPR) and providing technological support to suppliers in foreign markets who may in turn use this knowledge to supply ones competitors and/or move up the value chain to compete with their past customers in local or international markets. A further important factor of *hidden* cost, which has been showing a substantial impact over the past years, is the potential increase in labour costs (i.e. changes in the economic environment of the supplying country). Given fast rising labour costs in many developing countries, buyers, when negotiating follow-on contracts, may find that prices have risen steeply, requiring them to switch suppliers, leading to additional transaction costs.

The table below summarises the main elements of *static*, *dynamic* and *hidden* costs in global sourcing or offshoring that need to be considered for a total cost model.

In terms of key findings, we found three fundamental aspects that apply to offshoring and global sourcing decisions alike:

1. In the first instance, global sources scenarios will generally be of lower cost compared to domestic sources. The reason is predominantly the lower labour cost.
2. Dynamic costs, such as increased stock-holding and obsolescence, as well as the need for potentially expedited shipments (airfreight) generally does not tip this balance in the short-term.
3. What does generally tip the balance in favour of domestic sourcing, in the medium term, are the hidden costs related to strategic risks, such as currency fluctuations, labour cost inflation, and corruption.

It is vital for UK suppliers to make their business case not just on static costs ("unit cost each supplier gate"), but on a total supply chain cost basis, as otherwise their key competitive advantage will not be considered in the costing!

Static cost	Dynamic cost	Hidden cost
Purchase price ex factory gate	Increased pipeline and safety stock due, which is amplified by demand volatility and product variety	Labour cost inflation due to rising standards of living and competition o the labour market
Transportation cost per unit, assuming no unexpected delays or quality problems	Inventory obsolescence due to long logistics lead-times, e.g. in case of quality problems	Currency fluctuations, in particular for cases of artificially pegged currencies
Customs and duty to clear a shipment for export	Cost of lost sales and stock-outs, as the supply chain is unresponsive to shifts in demand	Rise in transportation cost, e.g. due to higher oil price and carbon offset costs
Insurance and transaction cost	Expedited shipments, e.g. air freight, to ensure uninterrupted supply	Overhead for managing the international supply base, including travel cost or cost for local personnel in the supplying markets
Cost of quality control and compliance with safety and environmental standards		The loss of intellectual property to contract manufacturers
Search cost and agency fees to identify and interact with local suppliers		The risk of political and economic instability or change

Source: Holweg et al. (forthcoming)

The traditional costing models (which largely consider static costs only) do not capture any cost savings related to the benefits of proximity between supplier and OEM. In that sense, UK suppliers will be disadvantaged from the start if they try to compete with low-cost countries (LCCs) on a unit cost basis.

Instead, a Total Supply Chain Cost (TSCC) model should be used, which is capable of capturing the strategic risks related to sourcing from LCCs, and from maintaining supply lines that span across the globe.

We have developed such a TSCC model in Excel, which is shown below. In general, there is always a trade-off in modelling: one the one hand the model aims to be as realistic as possible, on the other hand, the more complicated a model becomes, the harder it is to validate. In this case we have aimed for simplicity, for two reasons: firstly to ensure that the model could be used in practice, and secondly, so that it is easily understood and audited by the OEMs. Any model must clearly convinced all parties in the supply chain, and the more parameters and assumptions there are, the harder it will be to make a solid case for domestic sourcing. The model allows not only to make a judgement about current cost, more importantly, it also allows to predict future costs for either scenario, based on existing trends. For example, if one considers labour cost inflation and currency risk in the comparison of domestic versus global sources, one soon realises that these cost curves will meet eventually.

Total Supply Chain Cost Model

Description		Offshore	%	Domestic	%
Cost of Unit (ex factory)	c	42		50	
lead time, weeks	L	8		1	
std. dev. Of LT	sigmaL	1.5		0.5	
unit transportation cost	t	1.5		0.1	
value of unit for inventory	v	43.5		50.1	
annual inventory carrying cost fraction	I	0.1		0.1	
annual demand	D	11310		11310	
weekly demand	mu	217.5		217.5	
coeff of variation, weekly dem	delta	0.229885057		0.229885057	
shortage cost per occurrence	b	0		0	
shortage cost per unit	B	0		0	
order quantity, in wks of demand	Q	4		1	
safety stock factor	k	2		2	
annual obsolescence factor	Deltap	0.1		0.1	
disruption safety stock	delta k	1		1	
freq of disruption					
cost of expedited shipment					
Cost Calculations					
shortage prob (std normal)		0.0228		0.0228	
Exp Shortage amt (stdzd)		0.0085		0.0085	
std deviation of demand over L		355.5827		119.6936	
Pipeline inventory		1740.0000		217.5000	
Cycle Stock		435.0000		108.7500	
Safety Stock		711.1654		239.3872	
Expected Shortage amt		37.7393		50.8141	
Exp. # shortage occurrences		0.2844		1.1375	
Cost Elements					
Prod + shipping cost		£491,985.00	94.9	£566,631.00	98.9
Inventory Costs					
Pipeline		£7,569.00	1.5	£1,089.68	0.2
Cycle Stock		£1,892.25	0.4	£544.84	0.1
Safety Stock		£3,093.57	0.6	£1,199.33	0.2
annual shortage Cost		£0.00	0.0	£0.00	0.0
Obsolescence, value erosion		£12,554.82	2.4	£2,833.84	0.5
Disruption cost		£1,546.78	0.3	£599.67	0.1
Total Annual Cost		**£518,641.42**	100.0	**£572,898.35**	100.0
Difference in Total Cost		-£54,256.93			
What LT drives Delta to 0?					

In other words, global sourcing will generally be the lower-cost solution the first place, yet due to higher variability and risk in global sourcing settings the costs will eventually even out. We call this the "indifference points", where domestic and global sourcing costs are identical. In our view these indifference points will occur in **all** sourcing scenarios, yet we do not expect to see any general patterns as to *when* this cost equality will occur. The timing of indifference is related to many factors, predominantly the labour content as a fraction of product cost, the volatility in labour and transportation cost, as well as general risk related to currency exchange rates, and the like.

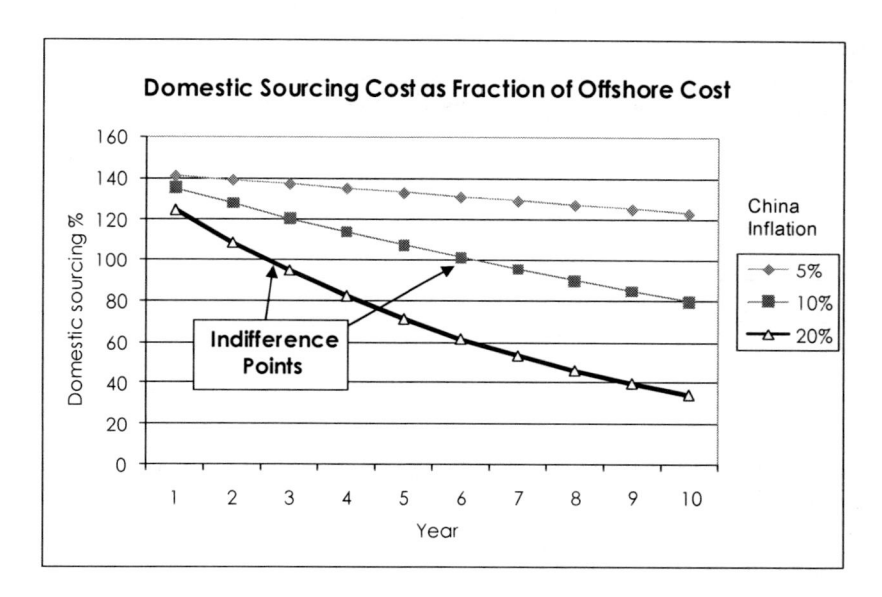

Domestic Sourcing Cost as Fraction of Offshore Cost

Chart showing domestic sourcing % over 10 years at China Inflation rates of 5%, 10%, and 20%, with Indifference Points marked.

Hence, the question is not *whether* there will be an indifference point, but *when*. Offshoring is almost *always* economically attractive in the first place, yet most firms that have offshored, do find that they are unable to realise the anticipated cost savings. The reason is the higher degree of risk and uncertainty in global supply chains, which turn will lead to indifference points between domestic and global sourcing.

The key opportunity for UK suppliers is to quantify the uncertainty and risk, and to build these factors into their costing models. Only if a total cost model of the supply chain is used will UK suppliers be able to present their key advantage in monetary terms: proximity!

(The Excel spreadsheet for the *Total Supply Chain Cost Model*, as well related working papers, are available free of charge from the authors. Please email Dr Matthias Holweg at the University of Cambridge for more information: m.holweg@jbs.cam.ac.uk).

References

BIS (2010), *Manufacturing in the UK: An Economic Analysis of the Sector*, BIS Economics Paper No. 10A, December 2010

de Treville, S. and Trigeorgis, L. (2010) 'It may be cheaper to manufacture at home'. *Harvard Business Review*, October 2010

Holweg, M., Davies, P. and Podpolny, D. (2009) *The competitive status of the UK automotive industry*. Buckingham: PICSIE Books.

Holweg, M., Reichhart, A. and Hong, E. (2011) 'On risk and cost in global sourcing.' *International Journal of Production Economics* (DOI: 10.1016/j.ijpe.2010.04.003) (forthcoming)

McKinsey (2008) *Time to rethink offshoring?* McKinsey Quarterly, September 2008

New Automotive Innovation and Growth Team (2009) *An independent report on the future of the UK automotive industry*, Department for Business Enterprise and Regulatory Reform, BERR/Pub 8860/0.5K/05/09/NP

ENDNOTES

[i] Holweg, M., Davies, P. and Podpolny, D. (2009) *The competitive status of the UK automotive industry*. Buckingham: PICSIE Books.

[ii] Ibid

[iii] Source: ONS, ABI 2009 data (latest available)

[iv] Source: ONS, BRES data

[v] SIC is the Standard Industrial Classification. A new classification was introduced in 2007 but for the automotive sector there is almost direct read-across, so continuity has been maintained.

[vi] Source: ONS, ABI 2009 data

[vii] GVA measures the contribution to the economy of each individual producer, industry or sector in the United Kingdom. It is defined as the difference between output and intermediate consumption for any given sector/industry. That is the difference between the value of goods and services produced and the cost of raw materials and other inputs which are used up in production.

[viii] Source: ONS, Trade in Goods, MQ10

[ix] Source: ONS, IDBR 2010. Figures are for SIC29 (Vehicles and parts) as no subdivision is available to separate out supply chain companies

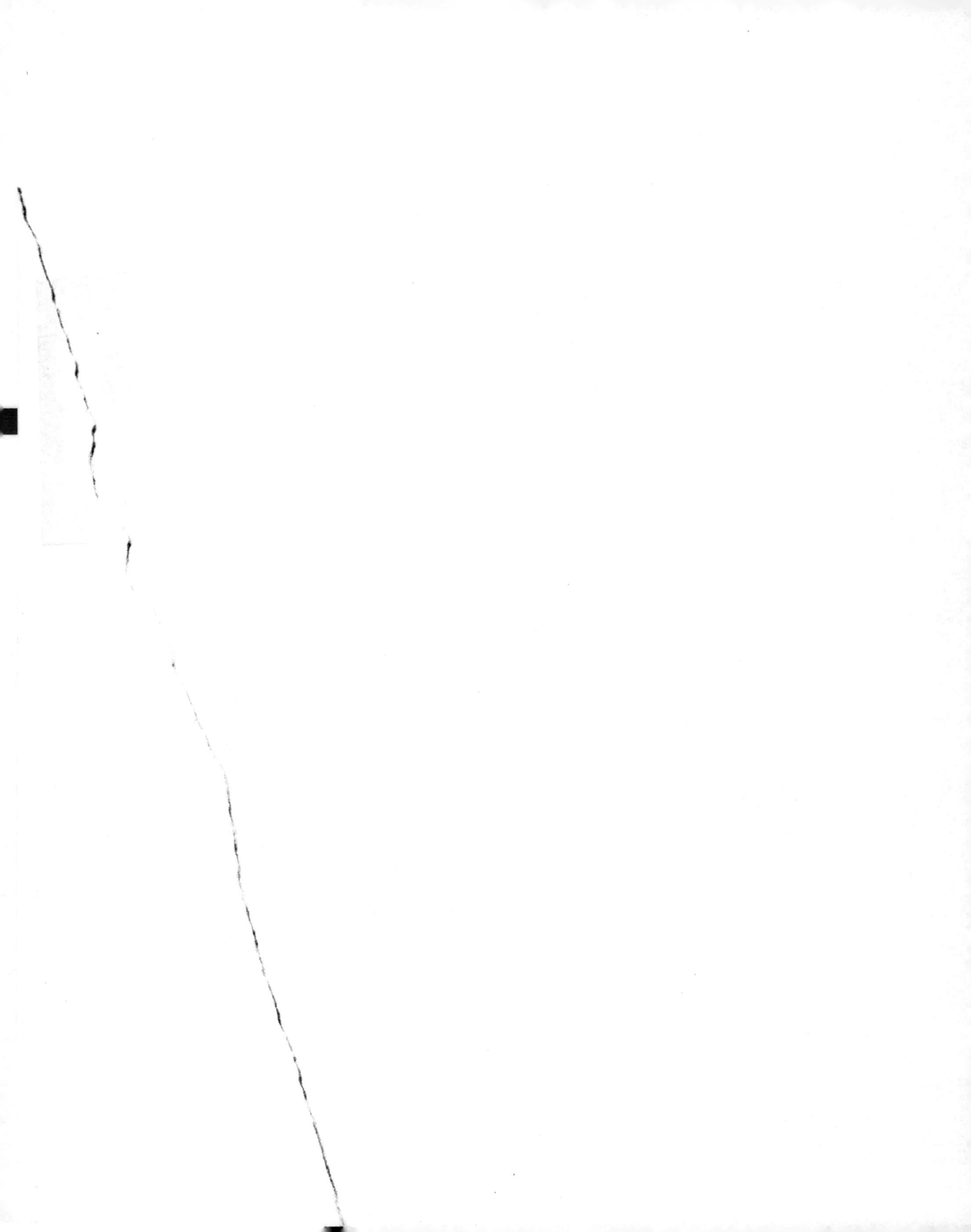

Lightning Source UK Ltd.
Milton Keynes UK
UKOW040307050212

186698UK00001B/5/P

9 780954 124496